The Art Of Network Migrations And Mergers For Service Providers

René F. Santiago

Copyright © René Santiago, 2022

All rights reserved.

No part of this publication may be copied or reproduced in any format by any means, electronic or otherwise, without prior written consent from the copyright owner and publisher.

However, quoting for reviews or educational purposes or for books, videos, or articles is encouraged and requires no compensation nor a request to or from the author or publisher.

René Santiago

About the Author

René F. Santiago is a network architect who has been working on the design, implementation, architecture, operation, and troubleshooting of Service Provider networks for almost 15 years. René is a Cisco Certified Design Expert (#20210013) and a Cisco Certified Internetwork Expert Enterprise Infrastructure (#23293). René spent many years at Cisco as a consultant for small, medium, and large Service Provider customers. He consulted on many technologies, including BGP, IGP, MPLS/VPN, MPLS TE, Security, QoS, and many other Service Provider technologies. René is now serving as the network architect for a top ten Multi-System Operator in the United States. René's previous publication includes the technical white paper *Traffic Redirect to Centralized Location with PBR over an MPLS Backbone*.

Dedication

Life is like the evolution of a network, where you try to adapt, improve, strengthen, and take care of yourself while enjoying the fast ride. This evolution has been shaped by multiple important people and places in my life.

Out of all of those, I am dedicating this book to my parents Nivea Maymí and Eric F. Santiago as they not only gave me life but also an immense amount of love to shape my evolution. Even to this day, they keep pushing to ensure I reach my full potential as a human being. They taught me to live life to the fullest while always helping others as much as I could. They taught me to always continue learning, no matter life's circumstances.

"Thank you" is not sufficient to describe my gratitude towards them and all they have done for me. For they have made tough decisions for me, like letting me leave my country to go to college thousands of miles away from them. And for showing me that there are no limitations to what goals you put on yourself—the only limitation is how hard you are willing to work to attain them. I hope this book makes them proud.

Acknowledgments

Thanks to my girlfriend Alexandra Rivera, for always supporting the challenges I put on myself and encouraging me to push through them. For giving me the space and time to write this book and loving me endlessly throughout the process.

Thanks to my English teacher Virginia Baéz from my high school St. Mary's School, who taught me how to write in English properly and with confidence even though it was our second language.

Thanks to my professor Aura Ganz from the University of Massachusetts-Amherst. She recognized my deep interest in computer networks before I even knew it. Thank you for guiding me down this path.

Thanks to my big brother Ricard Badia, the one whom I always followed since I was young. First with a love for mathematics and then with a love for network engineering. I will always be in his debt.

Thanks to my sister Carla Santiago for watching and caring for me my entire life. Thank you for giving me the confidence to push past all obstacles.

Thanks to my friend Ramiro Garza for providing amazing feedback on this book.

Finally, thanks to my country Puerto Rico. My desire to make you proud has been a true inspiration to write this book.

Introduction

Coming into the Service Provider network engineering industry was not an easy task. You may get overwhelmed by the sheer size of the networks and their complexity. I've noticed that many engineers are willing and able to do higher-level changes and migration projects but lack the confidence to say yes to those because they feel unprepared. This book shows the end-to-end process to tackle complex migrations and mergers. It provides a step-by-step approach to the things to do and what to look out for to complete such projects. Therefore, providing a process template for all Service Provider engineers out there who want to learn how to properly do migrations and mergers.

Chapter 1, "Identifying the Goal," explains the importance of knowing the project's purpose and what information you must collect when the project comes to the engineer in charge.

Chapter 2, "Technical Planning," describes the initial stage of comparison between the current production deployment and the newly proposed solution. It also covers the pieces of information that should be documented.

Chapter 3, "Developing the High-level Migration Plan," goes over the thought process of identifying the overlaps between the legacy and the proposed solution and how that affects your plan. Also describes how the impact of a migration should be considered and how to integrate that into your plan.

Chapter 4, "Technical Migration Tools," is the semi-technical portion of the book where certain network engineering tools/features are highlighted due to their usefulness in a migration or merger.

Chapter 5, "Testing," details the setup to follow when testing a migration and how it is the starting point of your "Methods of Procedures" (MOPs). It covers the different levels of issues and what to do for each.

Chapter 6, "Low-level Migration Plan," is all about how to build the configurations and the order of steps that will be executed the night of each change window.

Chapter 7, "Executing the Migration," depicts how to select the starting point of your changes as well as when and why you should do a peer review of your MOPs.

Chapter 8, "Mergers," covers the different types of network mergers and how to plan for each. It covers topics like documentation, IPv4 conflicts, and technical caveats likely to be encountered under a network acquisition.

Chapter 9, "Migration Variables," sheds light on things that you could need for some migrations, but not all, such as ordering equipment, training, external resources, and "magic."

Table of Contents

CHAPTER 1: IDENTIFYING THE GOAL 1
   New Project: Where to Start? *1*
   Requirements *3*

CHAPTER 2: TECHNICAL PLANNING 7
   Design Solution *7*
   Technical Documentation *8*

CHAPTER 3: DEVELOPING THE HIGH-LEVEL MIGRATION PLAN.13
   Getting Started *13*
   Solution Migration *14*
   Hardware Refreshes *18*
   Transforming Architectures *21*
   Impact *23*
   Underlay versus Overlay *26*
   High-level Migration Order *26*

CHAPTER 4: TECHNICAL MIGRATION TOOLS 33
   Admin Distance *33*
   Link Costing *34*
   BGP Local-AS *35*
   BGP Attributes *37*
   Staging *39*
   Redistribution *39*

CHAPTER 5: TESTING 41
    Lab Setup *41*
    Testing the Migration Plan *43*
    Issues, Bugs, and Workarounds *44*
    Configuration Templates & Standardization *45*
    No Lab Available *47*

CHAPTER 6: LOW-LEVEL MIGRATION PLAN 49
    Detailing the Steps *49*
    Building the MOPs *52*

CHAPTER 7: EXECUTING THE MIGRATION 55
    Setting the Order *55*
    Peer Review *56*
    Implementation *58*

CHAPTER 8: MERGERS 61
    Access *61*
    Documentation *64*
    IP Planning *64*
    Project Tracking *66*
    Merger Technical Plan *67*
    Technical Caveats *74*

CHAPTER 9: MIGRATION VARIABLES 81
    Equipment Ordering *81*
    Resources *82*
    Team Training *85*
    Emergency Changes *85*
    Creative Network Engineering *86*

INDEX 89

Chapter 1
Identifying the Goal

New Project: Where to Start?

Don't get overwhelmed. Change is the name of the game in any technical industry. Just because you got a new project that may seem like unknown territory or too big and complicated for you, stay calm and go through the guidelines stated in this book to make it a successful project.

In a Service Provider (SP) network, changes are constant. Whether this means bandwidth upgrades, new service offerings, special customer requests, mergers, hardware refreshes, software upgrades, etc. The first step in this process is to identify what the end goal is. What are we trying to achieve as a business and what value will it bring either to external or internal customers?

A project can be dropped on you or created by you. Usually, when a project gets assigned to you, it means the leadership either trusts you on executing it successfully or they are giving you an opportunity to show your skills. Ask many questions about the project. We will go over some of these in the upcoming sections, but this is highlighted as you mustn't stay quiet and accept general descriptions. You need to collect as much detail as possible.

If you are the one creating the project, you should prepare a brief presentation to the higher-ups that highlights the what, why, and how of the migration. You should list the benefits the project will bring and the high-level phases to finish it. At this point, you are looking for the higher-ups to give you the go-ahead on the high-level design. You don't want to get too technical yet. Just focus on the benefits the migration will bring to the customers, the company, or both.

No matter if the project is assigned or created by you, at the end of this discovery phase, you should have a clear understanding of the following:

1. What are the purposes of the project?
2. What benefits will this bring to customer experience, the business, and/or productivity?
3. What are the roles of my team if this is a collaboration project with other departments?
4. What are the timelines of the project?

Yes, this is not what your technical side of the brain expects for the first part of this book. However, having these objectives clearly defined first will be of immense benefit for the completion of the project.

One important note that deserves attention. Sometimes you might be brought into a meeting with the Vice President (VP) or even the C-level team of your company at the start of a new project. Do not be shy in expressing any concerns you may have. You are in the meeting for a reason. Some reasons that could make the project not feasible are: the timelines aren't achievable, there is no current solution for what they are trying to achieve, or you don't think you have the right team for it. Any reason that may be a red flag for you, that can affect the project purpose or delivery, this is the time to express them and address them. The leadership team would appreciate the input.

To offer a positive thought and not scare you away from further reading this book, have fun. New projects are exciting. They mean the network is evolving. They mean there is work to be done, which means job security. They also mean you can have fun completing a challenge. Be organized and thoughtful, but have fun with the process.

Requirements

With every migration, there should be at least one business objective to be achieved. This could be an external or internal business objective, and in some cases a mixture of both. An external business objective is something that will bring some value to your customers, which in the SP space are residential subscribers or business customers. An internal objective will improve the way the company does something. For example, migrating to a new management network architecture, an internal Border Gateway Protocol (BGP) architecture, and many other examples.

To achieve these objectives, there will be requirements. The same as with the objectives, there are business and technical requirements. You should have a list stating each and any concerns about meeting any of them. Some business requirements may be:

a. Ensure customers with unpaid bills are redirected to billing with services deactivated
b. Scale the commercial services architecture with a better Return on Investment (ROI)
c. Minimize IPv4 address purchasing
d. Enhance the services portfolio

I only named a few to give you an idea of the wording you may hear when you start a project. You should keep in mind the different factors that can affect your migration design and planning. What factors of the business are these requirements focused on? Is the business trying to lower Operating Expenses (OPEX) or Capital Expenditures (CAPEX)? Is the only focus increasing ROI? Is there a budget for the project? If so, who owns it? These are some questions you should have clear answers to.

Business requirements sometimes may be vague. It is your job to get clarification on them and what the actual use case is that they are trying to achieve. Only then can you move on to the next part, technical requirements.

You may not be super interested in business requirements. A lot of engineers just want to do engineering. As a technical lead though, you won't be doing your job as effectively as possible if you don't understand the business requirements. At the end of the day, the migration needs to align the technical requirements to the business objectives. Otherwise, you may put a solution out there that won't survive the test of time or that might be unintentionally too complex to achieve the goal of the business.

You are probably aware of what technical requirements look like, but let's name one for each of the business requirements examples stated above to give you a good reference.

a. Ensure customers with unpaid bills are redirected to billing with services deactivated
Technical requirement: Must utilize the core network to get all customers to a centralized server

b. Scale the commercial services architecture with a better ROI

   Technical requirement: New design/architecture should minimize configuration touchpoints

c. Minimize IPv4 address purchasing

   Technical requirement: Solution must utilize the current network backbone; no dedicated network will be built for this

d. Enhance the services portfolio

   Technical requirement: Ethernet Virtual Private Network (EVPN) must be utilized for the new Layer 2 VPN (L2VPN) Multipoint service being added to the portfolio

Sometimes the business and technical requirements will come from different people. Make sure you clarify all doubts from all sides involved. Technical requirements are going to be your foundation to put a migration plan together and execute it. However, as many network designers will often repeat, keep it simple. Your design and migration plan should be as simple as possible to meet both the business and technical requirements. Do not add flashy features just because vendors are saying you should. Don't put a migration plan together integrating many moving parts if you can build a sort of greenfield solution if budget and time permits. These are things we will discuss in the upcoming sections in detail, but just remember these words—"keep it simple."

Chapter 2
Technical Planning

Design Solution

Even though this is not a technical design book, it is worth briefly talking about how the design drives and affects the migration plan. Depending on the project and the resources in your company, you may receive the design solution from someone else, or you might be the one responsible for designing the solution along with the migration plan.

No matter where the design came from, you will essentially have to merge two designs, the existing one and the one you are moving to. To achieve this, you should first have a clear understanding of the technologies and platforms involved in each of the designs. This is called the discovery process. It is called discovery because there is a long road between putting the design in theory to getting the design deployed in production.

At this point, you should identify all knowledge holes of the technologies and platforms of the design, especially when someone else puts the design together. You can look at formal training, books, YouTube videos, blogs, and others to fill the knowledge gap. Some engineers just don't take the time to know the fundamentals of each technology, but this is an important factor in being successful in the implementation. Also, once you start the migration, you will be one of the people operations will reach out to when something breaks. If you don't have a clear understanding of how the technology works, you will spend a lot of time troubleshooting along with a lot of stress from your peers and managers waiting on you to figure things out. Make sure you know what you are planning to put on the production network. This can't be stressed enough.

Technical Documentation

Now that you understand both designs, it is time to document them. Both design solutions should be drawn independently of each other. Usually, the best way is to first draw the current design deployed in production. This documentation might already exist, it depends on how good your company is at keeping documentation up to date.

Either way, put as much detail into the documentation as you deem necessary. The details will help you cover any area of concern during the migration. For example, let's say you have one router with two uplinks to a core router and both uplinks are port channels. You may decide to just draw the port channel as one link and not even note the aggregated bandwidth of each port channel. Let's also assume one of the port channel's aggregated bandwidth is just half of the other one. During the migration, let's say you have to shut down the port channel with the bigger bandwidth and your secondary port channel will be saturated and customer traffic would be dropped. You just created customer downtime because you didn't note the details at the early stages of the process. This is just one simple example, but there are countless others. Hence, put as many details as possible on your technical drawings.

Once you have the current design drawn, you can separately draw the proposed design. Again, this might have already been done by someone else. Make sure the details are covered well. Now you compare both solutions and start identifying and documenting areas of concern for the migration plan. Make sure you document the concerns in detail too, as we'll cover each of them later on in the testing phase of the process.

Below are areas of concern that you should be on the lookout for and should note in your process.

1. Software and hardware support
   Note all the hardware versions involved and what software they are running. You should take the time to

do an audit and determine if certain hardware doesn't support one of the features in the proposed design. This will save you time during the migration testing phase. Note the port density of each device. Sometimes it is worth upgrading during migration, but you need the proper hardware and software for it.

2. Bandwidth

    Depending on the proposed solution, traffic might be redirected somewhere else. Or maybe you plan to standardize the network as you migrate, which is recommended when possible if major disruptions can be avoided. Do any of the links need upgrading from 10G to 100G or 100G to 400G, or from single links to port channels? Do you have the hardware to support such changes?

3. Interoperability

    Are you introducing a new vendor into the network? Are there any known caveats of interop between the vendors? Do all vendors support the proposed features?

4. Single point of failures

    This one is self-explanatory but important. This not only covers routers but also links, line cards, gateways, centralized systems, controllers, etc. Anything that when migrated could cause access or traffic issues should be pointed out.

5. Network protocols

   Without going into technical details, you should list all the protocols involved. You should also determine if any new protocols are being introduced in the proposed design and if there are any known interoperability caveats with the protocols in the current design.

6. Geography

   Document the name and location of each remote site and to where and what headend they connect to. Also note the locations of any major Points of Presence (POPs), and transit/peering points. This will become handy when figuring out what optical connections, if any, are needed for the migration or merger and what is the feasibility of such connections given the timeline.

Now that you have documented the designs and have a better understanding of both the starting point and the end goal, we can deep dive into the steps to follow to develop the migration plan.

Chapter 3

Developing the High-level Migration Plan

Getting Started

Strategizing a migration plan is never easy. You must not only take into consideration the two designs but also the impact of your changes. Unlike an enterprise network, a service provider network has customer traffic on it 24/7 all days of the year, especially during these modern times when the Internet is heavily used all day. In essence, your main goal when starting the development of the migration is to minimize the impact as much as possible. Yes, getting to the final solution is important but the customers come first.

Take a close look at both designs and determine what would be your first step. Depending on the migration, your first step might be building the infrastructure. Infrastructure could mean many things. Maybe you need new devices, line cards, optical equipment, etc. It could also mean you need new space to install new racks along with the power to deploy the new architecture you are migrating to. The two main things that take time to complete in terms of infrastructure are transport buildout and equipment lead times from vendors. If optical transport is needed for the migration, get an idea of timelines from the optical team to align them with the network migration timelines of the project. Later in the book, we will go into more detail on equipment ordering and lead times. For now, keep in mind this is an important factor in terms of timelines.

If no infrastructure is needed or if you can work around it while it is being built, then you can start the high-level migration planning. Network migrations usually consist of moving between solutions, transforming architectures, hardware refreshes, or a combination of two or more of the previous ones mentioned. We will go over each of these cases as each one would require a different thought process when it comes to developing the migration plan.

Solution Migration

When migrating between two solutions, you need to first identify the main components of each. There are always multiple components to a solution, but most solutions have a set list of components that are the building blocks of the solution. You need to identify the main components to see how they interoperate with each other and if any correlations between them would require special configurations and migration steps.

Once you have identified the components, see if any of them is part of both solutions. There may be more than one component that overlaps. If you have any overlapping component(s), you should deep dive into its functionality if you don't know it well, but most importantly you should get to know how to manipulate it. By manipulating, I mean how you can configure it to prefer one solution over the other. The solution here could be routes, labels, traffic forwarding, or any other component that would be needed to be preferred over the legacy solution to have a successful migration. Now let's go into an example to explain how this process works.

Let's say you want to migrate your network from traditional IPv6 core routing to 6 Provider Edge (6PE) to utilize the Multi-Protocol Label Switching (MPLS) backbone. For regular IPv6 routing, the main components are a routing protocol to advertise and learn the IPv6 routes through the core, and, of course, IPv6 addressing on the network components (interfaces and loopbacks) you want on that network. There may be cases where the network uses BGP to distribute the IPv6 routes and the Interior Gateway Protocol (IGP) is only used to advertise the IPv6 loopbacks for BGP next-hop routing. For 6PE, the main components would be an IPv4 routing protocol to advertise the edge loopbacks, MPLS in the core, BGP to advertise a label for each IPv6 loopback, and, of course, IPv6 addressing but only on the loopbacks you want to reach.

Staying with this example, let's assume we are using BGP on both solutions. We now see this as a main component of both solutions and something we must keep in mind when putting the migration plan together. You must determine which route will BGP prefer if you deploy 6PE in parallel. What about the Route Reflectors (RRs), will they come into play, and how? Will you have to use local preference or some other tool to prefer the routes, or will you have to shut down the IPv6 unicast address family BGP session? These are a set of the types of questions you should ask yourself once you find the component(s) that overlap between the two solutions.

If there isn't a component that overlaps both solutions, then you can start thinking about the order of operations at a high level. Why high level? Because you must look at the migration from the 50,000-foot view before you put any configuration together. One of the first things you should ask yourself at this point is: what components of the new solution can you deploy without affecting any of the current services or functionalities of the production network? Start with the main components and see if any of those or all of them can be deployed in parallel to the legacy solution from the beginning, and then you can steer the actual traffic to the new solution seamlessly in the later steps of the process.

Always keep in mind *preference*. There might be main components that at first glance don't seem to overlap between the solutions and you might think they will not affect anything on the current deployment if deployed in parallel, but you must always remember *preference*. For example, Open Shortest Path First (OSPF) and Intermediate System to Intermediate System (ISIS) are two different protocols and could be the main components of your solutions. However, you have to remember that by default OSPF is preferred over ISIS. If you have ISIS in your production network and start deploying OSPF in parallel because you may have thought the components don't overlap, then you will probably see some issues during implementation as traffic will be steered towards the OSPF network.

As part of your exercise at these early stages, you must not only check the overlapping of the main components but also *preference*. Go through the list of the main components and check which one of your devices will be preferred by default if deployed at the same time. Keep in mind that these preferences might change between vendors. Different vendors might prefer a certain protocol or feature over another. You must do your due diligence in checking the documentation properly for each vendor. You will more than likely see any misbehavior during testing, but it is better if you catch these caveats early in the process in case you need to add, replace, or upgrade network equipment to work around them.

We have been focusing on the main components of the solutions as we are designing the high-level migration plan. The minor components are no less important but will be covered in more detail during the testing phase of the project.

Hardware Refreshes

As part of the evolution of a network, devices should be replaced every couple of years. Some network devices have a longer lifespan than others, but eventually they should be replaced or upgraded with better/newer components. Hardware refreshes could be due to products reaching their end of life, feature support, bandwidth upgrade needs, port density, reduction of operational expenses, or constantly failing equipment.

When replacing a network device with another, there are mainly two approaches. Either replacing the entire device in one night or gradually replacing it with multiple change windows. Usually, you will see a smoother migration and less downtime with the second approach. However, that doesn't mean it is always the recommended way to go.

Doing a hardware migration in one night could be your only option if you don't have the physical infrastructure (space and power) to turn up the new device in parallel with the one being replaced. If the function of the device being replaced is not critical to the network, then doing a one night change could be a suitable approach. A network device that is constantly failing and needs to be replaced quickly to avoid more downtime could also be a good candidate for a one night migration.

When replacing hardware, you should always try to get the new device connected to the production network before the migration begins. This connection may be permanent and if so, even better. However, even if it is not needed long-term, it is recommended to have it to get remote connectivity completed ahead of time and it will give you a path for traffic to flow when the migration begins. It doesn't need to be a single connection. You can build all the permanent connections ahead of the actual migration of traffic if you have the resources for them.

With these connections done, you will speed up the time of the migration since you will need to move fewer connections. Having these connections will also allow you to complete the migrations in multiple nights in a more efficient matter as you will not necessarily need to affect the uplinks/core-facing connections of the device being replaced. That device can continue to operate as usual, and you can focus your plan on moving everything behind that device into the new device which is already actively connected to the production network.

When it comes to determining the order of the moves of the physical links to the new device, you should consider the importance of the link, the amount of traffic on it, and the potential for complex troubleshooting.

A link's importance comes down to what is connected at the other end. For example, if at the other end it's an Internet transit provider, then that's a very important connection. If at the other end there is a Trivial File Transfer Protocol (TFTP) server that's only used occasionally to move internal files, then that's not an important connection. Highly important connections should be moved as early as possible in the high-level plan to give you time in the change window to troubleshoot any issues with such connections.

Links with larger amounts of traffic should be moved first as well. Note that just because a link has a small amount of traffic doesn't mean it is not highly important. For example, connections to the Data Center (DC) infrastructure where your company's website and provisioning servers are located might not have gigabits of traffic, but they are essential to the business and customers and this makes them highly important connections.

Finally, you should determine each link's level of complexity. For example, a link that is just a trunk is not as complex as a link that has a combination of multiple protocols like Protocol Independent Multicast (PIM), IGP, Bidirectional Forwarding Detection (BFD), etc. The more complexity a link has, the higher priority it should have to give yourself time for troubleshooting such a level of complexity.

Transforming Architectures

All networks evolve, some at a faster pace than others. This evolution happens via replacing architectures. Most of the time the new architecture is due to a business requirement but sometimes it is just a technical one. You may say that a new architecture could apply to the previous sections of "Solution Migration" and "Hardware Refreshes." While this is somewhat true, new architectures are generally a bit more involved since they could require both new solutions as well as new hardware. The architecture could be a complete overhaul of one design being replaced by another. Regardless of the scenario and reason for the new architecture, it will be your job to design the migration to it.

When migrating between architectures, you should first determine the components of the new architecture, and of those, document which ones are already in place in the production network. Now you will know what components are going to be new to the production network, and come up with an order to implement them without affecting the functionality of the traffic. If the components are new hardware, connecting and activating them into the production network should be your first order of business in the high-level plan.

Once the applicable hardware is in place, your next focus in the high-level plan should be building the control plane for the new architecture. You should first activate the components that will not impact production traffic. Always consider *preference* when doing these activations to avoid altering the traffic in the network. For the components that are already in the production network but will be utilized in the new architecture, you should answer the following question before putting their implementation in the high-level plan:

a. Are there components of the new architecture that will need to be redesigned? For example, multicast VPN (mVPN) Generic Routing Encapsulation (GRE) Rosen-draft requires VPNV4 and IPv4 Multicast Distribution Tree (MDT) address family BGP sessions. Will you use the same BGP architecture currently in the network, or will a new one need to be built?

b. Are there any prerequisites that need to be in place in the current network before the migration to the new architecture can take place? For example, let's say you are taking out "next-hop-self" from the Internet border routers and just advertising the transit links into OSPF via passive. You need to ensure that all devices in the "BGP-free MPLS network" are now running BGP because otherwise they will drop the traffic since they will not have labels to the transit link advertisements if you are just creating labels for host routes.

c. What and where are the temporary implementations in the current components that will affect the new architecture and will need to be adjusted? For example, let's say you are migrating to a single OSPF area in one of the segments of your network, but you have utilized the multi-area OSPF feature as a temporary implementation. If you don't remember to adjust this, you could see a traffic flow you weren't expecting due to OSPF's preference of inter-area routes.

These answers will be the key factors to determine the final steps of your high-level plan. As can be seen, migrating between architectures requires a deeper thought process compared to what was covered in the "Solution Migration" section. Sometimes the types of questions stated here are already answered by the network architect and sometimes it may be up to you to determine this information. You shouldn't build a high-level migration plan without this information.

Impact

In an SP environment, there is no such thing as no impact. SPs serve customers 24/7 all year round. However, there are times of less traffic activity in the network. Each SP must identify what those times are depending on their customers, the type of services, and geographical location. For example, a pure MPLS backbone serving only business customers point-to-point or multipoint will have higher traffic activity during regular business hours. On the other hand, an SP focused on residential customers will see higher traffic consumption during evening hours when people are home. You must identify those hours early on to know when you can execute a migration.

The impact on customers might be pretty straightforward during a migration. However, some customer services rely on systems that may be internal to the SP environment that could still affect customers. What this means is, don't assume other changes in the calendar during your change will not affect customers just because they are highlighted as internal or as system changes.

Let's take for example a Multi-System Operator (MSO) that provides the complete array of residential services (voice, video, and data). On the data side, let's say you are migrating from multiple single links to a port channel, and that same night the systems team is updating the Domain Name System (DNS) servers. When you are in the middle of your change and are troubleshooting something, you may not have realized that the reason your customers can't surf the Internet is not because of your routing change but more than likely because the DNS system is being worked on and there are no DNS servers available.

Going back to the same port channel example, let's assume there is a change that night on your voice provider systems. Again, you may be thinking there is a routing issue between your edge network and the voice provider, but it may very well be it coincided with the change on the voice provider side.

The point here is that you should coordinate accordingly across teams and keep an eye out for other changes in the network, either to the core network or systems. Sometimes you may also have to coordinate the changes with those internal teams to tackle multiple things in one night instead of affecting services on multiple nights. It all depends, but these are things that you must keep in the back of your mind when putting in place a migration plan.

This is highlighted here because it truly depends on the migration project. Some migration projects won't be impacted by other systems that much and can be either ignored or noticed during the testing phase. Other migration projects would have many variables that may need to change all at once and it is highly beneficial to identify these impacts as early as possible in the process. Later, we will talk about where and when to start your migration depending on the risk of impact on services.

Underlay versus Overlay

When talking about networking architectures, there are underlay and overlay components/protocols. Overlay network protocols ride over an underlay network, meaning the overlay can't run without the underlay pieces in place. It is important to identify if the new network solution you are migrating to has an underlay and an overlay as it will come into play in your high-level migration plan since you will need to ensure you put the build-out of the underlay first, before the overlay in your migration order.

The other important reason for identifying the underlay and overlay components on a migration project is to determine if you need to build both for the migration, or if you can use either the underlay or overlay components on the existing solution. The case may be that you may just need to make small changes to either component on the current solution to use them. These are all important factors to keep in mind when you put your high-level migration plan in place and hence a step you should take in the process before moving forward.

High-level Migration Order

Now that you have a better understanding of how the solutions interact with each other, you can start putting high-level steps together. At this time, you don't need to lay out the specifics. You just need to make sure your migration plan makes sense considering the caveats discussed in the previous section. You will get down to the nitty-gritty details in the testing phase.

To have a successful migration, you should always consider ordering your high-level steps from least impacting to most impacting. You start prepping the network slowly but surely until the grand finale. The order can be conservative or aggressive. Aggressive means combining steps or skipping steps that are there for safety measures. You can consider the aggressive approach when you meet at least one of the following criteria:

1. A very redundant network environment where you can shift traffic to one side of the network while you work on the other one.
2. Services are mainly used during a specific time window of the day which leaves you the rest of the time open for very low impact. This is highly unlikely on service provider networks, but possible.
3. Greenfield deployments. Networks that are being built brand new.
4. Migrations with very tight timelines. Sometimes you have to give away quality for execution. Even though that is never ideal, is it sometimes required.

Regardless of the approach, make sure your steps are clearly laid out and easy to understand. The reason for this is because your high-level plan will more than likely need to be presented to other teams or managers that are not network gurus but know enough about networking. These people may have a say in the decision process, so they must understand your plan and be satisfied it will be successful.

Let's look at an example to explain the thought process a bit better. Let's assume we are migrating business customers with Direct Internet Access (DIA) and/or point-to-point (P2P) services to a new platform under an MPLS architecture versus a Virtual Local Area Network (VLAN)-based architecture to provide the services. We can also assume the design and architecture have been given already and we are tasked with putting together the high-level migration plan.

Here is the high-level plan, and we will then discuss the thought process of why each step is in the order they have been put together.

1. Deploy the new network devices into the network
2. Identify each customer's service per location and their A and Z locations if they are P2P
3. Identify P2P customers that will have A location under the legacy architecture and the Z location on the new architecture
4. Enable routing and MPLS on the new architecture
5. Prepare configuration templates for each customer
6. Migrate DIA customers

7. Migrate P2P customers

First, notice how there are no details under each step even though there are many things that need to happen before completing each one. That is why it is called high-level. After the testing phase, we will get all the details and develop the actual Methods of Procedures (MOPs). This is the starting point of the project, and it should be completed regardless of the size and complexity of the project. It will ensure things run smoother and it minimizes the risk of missing key details and steps.

The first step is the infrastructure setup, and if you want, you could switch it with steps 3 and 4 as those steps are part of an audit that can be done before, during, or after the infrastructure setup. It is put there since it takes time to get it done due to vendor lead times for parts, rack space, power, software upgrades, onsite resources, etc. Therefore, we could be doing step 1 while another team is already working on step 2.

Steps 2 and 3 are part of the preparation audit process. Each customer should be listed, its network variables like VLANs, IP, location, speeds, service, etc. Also, since we are migrating between two architectures, we are not going to migrate all locations in one night (let's assume). Then there will be cases where one location is on MPLS and the other is still on VLAN-based architecture. You need to figure out a temporary solution to get these two locations talking. This must be identified ahead of time to properly prepare configs at both ends of the circuit for this temporary scenario.

Step 4 is one of those steps that can be completed without any impact if done properly, and it is an example of a step that may be done during regular work hours if your company allows it. "If done properly" because if you are not using automation, an engineer could fat finger the wrong loopback IP and input an IP already used in the network and routed. This could create an impact on services. It is recommended to get the routing activated before the migration of services to allow the new network gear to bake in the production network ahead of putting any customer traffic on it. If it is going to fail, let it fail during the baking period without much impact (hopefully).

From the audit process and after testing, you can build your configuration templates and apply them to each customer using their network variables that were documented in steps 2 and 3. You could use scripting for this, and it is highly recommended. You can prepare all the customer configurations, or you can do them in batches based on the order of the locations to be migrated. One thing to note is that you need to remember to replace the configs for any temporary solution, like the one discussed here in the previous paragraphs.

The last two steps are the actual migrations that would happen overnight. It is recommended to put the DIAs first because it is easier to understand and troubleshoot than an MPLS circuit. This way you can catch any major issues ahead of time. The Network Operations Center (NOC) will thank you for it when they get trouble calls from customers that were migrated but are still down since it will be easier for them to troubleshoot. We will discuss the number of changes per change window later, but keep in mind it is an important factor. Just because it says "migrate DIA customers" doesn't mean you will migrate all of them in one location on a single night.

Chapter 4
Technical Migration Tools

Even though this is not a technical book, it is important to highlight some of the tools you can use to complete your migration project. These tools are independent of each other but can be used together depending on the case scenario. These tools will help minimize the disruption of services in your migration and are tools to keep in mind when going into the testing phase of your migration before you develop your final MOPs.

Admin Distance

Routing protocol dynamic routes, static routes, and directly connected routes have an assigned admin distance. The admin distance is a number and the lower the admin distance, the more preferred that route is. If the same route is learned via different sources, the one with the lowest admin distance gets programmed into the routing table. This means you can have the same route learned via multiple ways at the same time on the same network, and only one will be used for traffic forwarding.

With this tool, you could build a routing protocol in parallel to the current routing protocol without affecting any of the traffic by making sure a higher admin distance is in place for that newly built routing protocol. Depending on the vendor support, you may be able to run different routing instances of the same routing protocol with altered admin distances. If that is not the case, some vendors will allow altering the admin distance for routes learned from specific sources. Either way, you could build an entire network end to end without much or any disruption to services. Later, in the migration execution process, you could shift traffic to your new network by either lowering the admin distance of the new protocol or increasing the admin distance of the existing routing protocol. The recommended approach is to keep the industry-standard admin distance on the new protocol and only alter the legacy routing protocol.

Link Costing

All routing protocols have a default cost per interface adjacency. This cost can be changed via configuration, and it is a tool that allows you to shift traffic effectively during a migration. You may shift traffic out of the link while you make a certain change on that path. You can also shift traffic into the link to migrate it from a legacy link. Finally, you can cost out all the links of a router to divert all traffic away from it while you complete a migration on it, perform a software upgrade or replace it with another device.

Regardless of the need and use case, it is always a best practice to remove the alteration as one of the final steps of the migration to keep the cost at the default value. This may not be the case on all migrations as you may need to keep the cost configuration for whatever reason, but, if possible, you want to take out any temporary changes from the network to keep its operation and troubleshooting as predictable as possible.

BGP Local-AS

Migrating BGP architectures can overwhelm some engineers. Rightfully so since BGP is one of the crucial and centerpiece components of most SP networks to distribute all sorts of different routes across it. Also, configuring a "no router bgp" statement is a bit nerve-racking. However, knowing the right tools and how to utilize them will make your BGP migrations much quicker and smoother.

BGP Local-AS is a tool that will help you when migrating between Autonomous System Numbers (ASNs). It is a feature that allows you to mask an ASN when building peering sessions with specific neighbors that require it. This is a feature that not all vendors support. Also, inside each vendor's portfolio, support could vary from platform to platform just like any other feature. The feature also has some extensions, depending on the vendor, that allows you to hide the original ASN from the AS path and others.

This tool comes in handy in migrations when, for example, you have three island networks all in their own ASN (1, 2, and 3). Let's assume you want to consolidate your three ASN islands into one ASN (ASN 1) to have better operational flexibility. We can go into ASN2 and configure the first core router with ASN 1 and we can use BGP Local-AS 2 to trick the neighbors still in ASN 2 to think they are still talking to a router on that same ASN. Our router is on ASN 1, but we are announcing to the legacy neighbors that we are still on ASN 2. This way traffic can continue to flow as expected while you go around the network migrating the routers in the same fashion.

Once you are done, you can remove the BGP Local-AS configs between migrated routers, and they would form an adjacency on ASN 1 only. This would allow you to split the changes into multiple nights since you will not have to build a new BGP architecture when you move to a new ASN in one or two routers. The other important note is that this is applied per neighbor. You can pick and choose which adjacency you want to mask and which you want to move on from the legacy ASN. It is all about being flexible with this tool and giving you more time to complete the BGP migration.

Another great benefit of the tool is that it can be utilized on only one side of the adjacency and your neighbor doesn't even notice it. This is extremely useful when you are migrating ASNs and have Internet transit providers, or private peering and/or caches with content providers where you will not have control of the remote end's BGP configuration. Another use case would be for network mergers after an acquisition. Just know this is a very useful tool that can be utilized for BGP migrations regardless of the address family being utilized.

BGP Attributes

As you may know, BGP uses path attributes for its best path calculation. Going over all the path attributes and the order in which BGP prefers them is outside the scope of this book. Let's go over the two most utilized ones in the SP environment, and how to manipulate them for a migration.

Local preference is a BGP path attribute that is applied to inbound BGP routes to alter outbound traffic. If you are migrating a BGP address family between two different BGP architectures, and you want to divert outbound traffic from the legacy architecture to the new one in a seamless and phased manner, then you can either lower the local preference of the legacy routes or increase the local preference of the new routes. You must make sure you know the current local preference of both architectures to properly adjust the configuration. The higher the local preference, the more the route is preferred.

AS path prepending is a BGP path attribute that is applied to outbound BGP routes to alter inbound traffic. If a path from two different sources is going through the AS path list comparison in BGP, the path with the least amount of ASNs in its AS path list will be preferred. Let's say you are migrating between two different Internet Transit Providers. You have the option to just shut down the links to the legacy provider and divert the traffic that way to the new provider. However, you can achieve the same results seamlessly by adding your ASN multiple times to the AS path list of your routes being advertised out the legacy provider and letting those propagate through the Internet. Now your traffic should be entering your network through the new provider as the Internet will prefer the shorter AS path list of your routes if all attributes before it are the same.

One of the aspects that make these tools useful is the fact that they are applied to routing policies, and you can alter traffic for specific routes only or for a lot of them. This allows you to phase out the BGP migration and minimize the risk of the impact of your changes. Also, both tools can be used for internal BGP (iBGP) or external BGP (eBGP) neighbors, which makes them highly flexible for migrations. Remember, these are useful tools for any BGP address family. Please keep in mind that there is a specific order BGP follows for its best path calculation, and the path attributes higher up in the order before local preference and as-path list need to be the same for the routes in question for these tools to be utilized properly for a migration.

Staging

Staging is the process of preparing network equipment and/or configurations in a predefined location before introducing it to the production network. The staging can be done while the device is connected to the production network or not. That all depends on whether or not you need remote access and don't have tools such as a console server.

This is a tool that can help expedite migrations for two reasons. One, you can have resources prepare the equipment and/or configurations while other things are being taken care of for the migration. Two, because the equipment is already prepped and configured, the deployment and disruption time on the actual migration window can be reduced significantly. A couple of the common things that can be done on a staging process for networks are software upgrades, applying configuration templates, applying IP addresses, test optics, and running reload or switchover tests.

Redistribution

Even though many network engineers are not true fans of redistribution because of its many risks, it is an invaluable tool for certain migration projects. When you have two or more separate routing networks and you need temporary connectivity between them for migration purposes, you can use redistribution. Also, it may be needed in some scenarios where the vendor software doesn't support any other way of inserting routes into a routing process other than redistribution.

If you are going to use redistribution for your migration project, you should keep in mind a few things to minimize the risk of disruption of services. You should document how many routing processes are on each network, and if there is redistribution between any of them. If there is redistribution, where are the redistribution points? Are they redistributing both ways or one way? Are they filtering just certain routes or all routes? This will save you hours of troubleshooting down the line.

You should always use some sort of filter (route map or route policy) when using redistribution instead of leaving it wide open. This way you protect your network from fat-fingering mistakes and unknown routing problems. You want the network to be as predictable as possible during any migration project. Please remember the best practice is to use redistribution only if necessary and to always remove it after you are done with it. Routing loops and unknown traffic patterns can creep up on you if you leave them out there.

Chapter 5
Testing

As with any technical networking project, testing should be done as a proof of concept. The theory is one thing but is not until you get your hands dirty that you can validate your theory and hash out all the configuration details of the migration. Also, you must know the migration from a technical standpoint to explain it properly to other members of your team or to other teams that would need to be involved in the project.

Lab Setup

When it comes to the network lab, the ideal scenario is to have a replica of the migration network(s) to ensure all components are tested. However, this sometimes may be extremely difficult because some of the equipment on the migration project may be very old and you may not have any available nor will the vendor be able to find any for you. Sometimes they may have a virtual appliance you can deploy on a virtual machine in the lab that you can test. Physical network equipment is always preferred because you want to test the hardware as well as the software. It may be the case that your only option is to set a network emulator as your lab and hopefully you can find the same network images as the components involved in the migration project.

If you can't have all the network components that are part of the migration in the lab, you should try to at least have one of each of the most critical components. The less important components can be replaced in the lab by another model or another similar version of that device. Once you have the components ready and on the right software version, try to mimic the physical connections as well. Be as detailed as possible, if you have single links do single links, if you have port channels then do port channels. Also, try to use the same type of port as in the production networks (1G, 10G, 25G, 40G, 100G, etc).

If the lab has a traffic simulator, that is ideal. There are several vendors out there that allow you to mimic production traffic closely, not only in the amount of traffic but also in the type of traffic. This would be essential as there are bugs out there on vendor equipment that may only show themselves when certain traffic triggers them. If you don't have a traffic simulator, that is not an issue. You can utilize regular ping test tools or some traffic generator test sets without having to simulate specific traffic patterns.

Once the physical lab is set up along with connections, you should copy all the relevant configs from the production network into your lab. You want to make sure you catch any software bugs and/or caveats while going into migration testing. When you have the lab setup mimicking the production network to your satisfaction, you can then move on to the actual execution testing of the migration plan.

Testing the Migration Plan

At this point, you should have your high-level migration plan in place. You have a theory or a vision of how you want to execute the migration. When it comes to testing, you want to follow the high-level migration plan steps. You shouldn't skip a step on the plan just to get items completed and checked off. The more realistic you keep the migration testing, the more successful you will be. This is because you will be able to identify how configurations from prior steps affect behavior on upcoming steps and vice versa.

When you execute each step of the migration, make sure to document a template of the configurations done for each. This is how you will start building the low-level migration plan where all the detailed configurations will be found. Having a configuration template for each step will also allow you to replicate them easily if that step needs to be implemented on multiple devices. It will also allow you to automate the configurations if you have the resources and abilities for that.

Issues, Bugs, and Workarounds

You will probably face several issues with your migration plan testing. The first thing you must do is document the issue and determine if it is what we call a showstopper or not. A showstopper is an issue where a major redesign of the migration plan is required. If the issue is a showstopper, make sure you communicate it to the team and start thinking of a possible redesign. You must decide whether it is worth it to redesign that specific step or component or whether it is best to start from scratch with a new migration plan.

If the issue is a minor thing, then look for a workaround that would satisfy the migration requirements without too many service disruptions or major changes to the migration plan. Some of these minor issues could be due to software or hardware support. Figure out what needs to be done and make the necessary upgrades. Other minor issues might have multiple fixes or workarounds. Always remember, you want to keep your migration as simple as possible from the technical side while fulfilling the requirements. This includes the time of implementation, amount of configuration, resources needed, and complexity of the solution.

No matter whether the issue is a showstopper or not, you shouldn't move forward past the step where you found the issue on the migration plan. You don't know how the fix for this issue can affect the prior or the following steps. It is hard to stop progress on testing, but, when it comes to issues, it should be done. It will pay dividends on your execution.

If you find a software bug in your testing, first identify its impact on your plan. Software bugs can be categorized from critical to cosmetic. You must identify its level of impact to determine how much pressure and how often you can or should apply to the vendor of the software. Sometimes bug fixes are already fixed in a later release that already exists and a software upgrade will resolve it.

If the bug is of minor impact to your plan and you can come up with a workaround, then great. If the bug is critical and there is no workaround, then this falls into the showstopper scenario. Let's assume there is no possible technical option or design to get around the bug. If that is the case, you must wait for the vendor to fix it. Bug fixes take time but if you have enough leverage to push the vendor, they can always get the fix done quicker and get an engineering release to you in a decent amount of time. Another option if you are in this spot, is to find another vendor.

Configuration Templates & Standardization

During every step of the migration plan, you should document all the configurations involved in each step. This should be done for all the devices involved in each step.

The configurations don't need to contain the specifics like names, descriptions, IPs, etc. They should just be the overall configuration templates that you can apply repeatedly on the devices.

At this point of the process, you can also start making decisions on one key network aspect, standardization. Standardization in networking refers to utilizing the same implementation on all the devices serving similar functions or in the same region. There are two separate ways of standardization that apply here. One is standardizing for a migration project, and the other is taking advantage of the downtime the migration will create to standardize other aspects of the network not directly involved with the project.

Some examples of standardizing as part of a migration project are splitting IP subnets for specific functions and geographical locations, using the same names for route policies, and utilizing the same values for each feature.

Standardization helps in many aspects of networking. It makes troubleshooting much easier as engineers can quickly identify the purpose of something by looking at its implementation. Therefore, reducing the time of investigation into an issue. It also makes automation much easier as you know exactly what the standard is and can code it in. Finally, it reduces the time of deployment even without automation, as you can follow the standard template and get it implemented without thinking if and where configurations would go and for what purpose.

No Lab Available

Not all companies have a lab set up to utilize for migration testing or for any testing for that matter. In this case, the recommendation is to find as many ways as possible to simplify your migration plan while minimizing the risk of outages. For example, if you just need to shift traffic to a new network and don't know if building a parallel IGP would break something, you could always deploy static routes across the network. Of course, this is an extreme example, but the idea is to highlight the simplicity of the deployment.

Because you won't be able to test your migration plan before executing it, it is recommended that you break down your low-level plan into as many steps as possible with backout procedures for each. This way if you find an issue on your plan, you know exactly which step you found it on. It will allow you to laser focus on the problem at hand while at the same time, expediting a fix or workaround for it.

This scenario also means you will be going somewhat blind to the migration. You should communicate this to the NOC and the rest of your team. An engineer from your team should be on the change with you at least for the first couple until you are confident you have found all the major issues and have fixed them. You should have an engineer from the NOC team on the changes as well to monitor the rest of the network while you do them. They will be your eyes and ears during the changes. They are an invaluable piece, especially in this scenario. Finally, you should coordinate to have an engineer from your team rested and ready to jump on a troubleshooting call the next morning. It is hard to keep focused for long hours especially when you start working at 1:00 a.m. and it is now 7:00 a.m.

Chapter 6
Low-level Migration Plan

You are now ready to start building the MOPs and start scheduling the change windows. This is all part of the low-level migration plan which comes right before the execution of it. This is the part of the process where all the details are put in for every single change. It is the most time-consuming part of the migration project but with some scripting, it can be completed much quicker.

Detailing the Steps

In the high-level plan, we had the overall steps of the strategy to follow. However, inside each of those steps, several steps need to be completed to complete that phase. These are the steps you need to detail in the low-level plan. Let's take two steps from the high-level migration plan used as an example in Chapter 3.

(Step 1 from Chapter 3) Deploy the new network devices into the network

1. Upgrade the software of the equipment and put a configuration template on it at staging
2. Rack, stack, and power the equipment
3. Run connections to the network and bring them up/up

(Step 4 from Chapter 3) Enable routing and MPLS on the new architecture

a. Ensure you can ping both ends of the link
b. Configure MTU to match at both ends
c. Add a high cost to the link on both ends
d. Turn up the IGP routing for the link in question at both ends
e. If required, configure authentication at both ends
f. Verify IGP neighbor session comes up at both ends
g. Advertise the loopback of the new device into the routing process
h. Ensure connectivity to the loopback of the new device from the rest of the network
i. Configure route policies with deny alls on the route reflectors as well as on the new device
j. Bring up the BGP session between the route reflectors and the new device
k. Verify the BGP session came up and neither side is advertising nor receiving routes
l. Enable MPLS Label Distribution Protocol (LDP) on both sides of the link
m. Verify the MPLS LDP session is up, and that you are creating and receiving labels

n. Remove high IGP costs at both ends of the link

Depending on the high-level migration plan step, you may see a few or many individual steps. There is no minimum or a maximum number of steps in a low-level migration plan. For complex migrations, it is a good idea to break them down into as many steps as possible to maximize the efficiency of completion. It is important to always put the steps in sequential order. If a step has a prerequisite to be completed, make sure that prerequisite is noted as a step before it is on the plan. For example, you need the underlay network done first before you can turn up the overlay network.

Adding the verifications as a separate step is crucial to the overall execution of the migration. Even though it is sort of obvious to some engineers that verifications must be done, it may not be obvious to others. If the verification step is not highlighted and is skipped by the implementation engineer, you could see some unforeseen issues and outages. Therefore, it is always best to note them on the low-level plan to check your previous work.

The verification steps also become rollback checkpoints. At this step, you can determine if everything is working as expected or if the changes need to be rolled back because you are seeing issues. You can verify steps after every single change, but the rule of thumb is to add them after a major milestone has been achieved. Especially a milestone that you know has the potential to disrupt services. Use them, they are invaluable to the success of your migration.

Building the MOPs

Once the low-level migration plan is completed and verified, you can start building the individual MOPs for each device on every step of the low-level migration plan. The MOP contains all the specific configurations for each step. The MOP building process can be completed quickly if you followed all the previous steps and have documented the configurations properly for each step based on your testing.

To build the configurations, you simply copy and paste from the template built during testing and fill out the missing parts like IP addresses, descriptions, router IDs, etc. This could be expedited using scripts, which are highly recommended for accelerating the process and also reducing fat-fingering mistakes. You could also include reference information relevant to those steps that were discovered during the audit process if one was completed. This will expedite any troubleshooting time for the implementation engineer.

All the steps should be divided into their section and accompanied by a short description. This is done to help the engineering team in case they need to review your MOP after you are done, and they are working on an issue caused by the change. It will help narrow down the possible configuration that caused the issue. One MOP should include multiple steps, but the recommendation is to terminate the MOP at a major milestone. For example, if you are building the routing for three devices on the same night, each device should get its MOP. This will keep your changes organized and documentation easier to find in case it is needed.

Verification is an integral part of the MOPs. It is highly suggested you have a verification procedure for every step on your MOP. Some more senior engineers usually do not highlight the exact verification commands utilized for every single step as they already know what and how to check the changes. However, for more junior engineers, it is recommended they state the exact verifications they will perform on their MOPs. These verifications should also be saved or logged to a file. This will not only help them expedite troubleshooting in case they run into any issues, but they will also learn how the changes they are making affect the network.

All your MOPs should have a backout plan, also known as a rollback procedure. The rollback procedure should be highlighted after specific sections on the MOPs. The recommendation is to put them after parts of the MOPs where traffic disruptions could be seen or after multiple significant steps. If you have only one rollback procedure at the end of each MOP, then it will take you more time to get back to the original configuration and revert the changes. You shouldn't wait until the last minute of the change window to perform configuration changes. Otherwise, you will have no time to perform your rollback procedure(s).

Chapter 7

Executing the Migration

Setting the Order

Once the configurations are completed, it is time to get into executing those MOPs. You must pick a starting point in the network where you will first start deploying the changes. The typical choice is to pick a portion of the network with the least number of customers and/or traffic on it. This is to minimize the impact of the change since it is your first time executing a change for the migration project in the production network. However, that is not the case for all migration projects. If you are a consultant for an SP, the SP can also help decide which part of their network is less critical.

Let's say for example that the project is to software upgrade your Provider Edge (PE) devices. Let's also assume that not all your PEs have the same set of services on them. Some may just have DIA customers on it, others may have a combination of DIA and P2P customers, and yet others may have DIAs, BGP customers, L2VPNs, and Layer 3 VPNs (L3VPN). For such a scenario, it is sometimes best to pick a PE that has all the services the company offers to see the code's behavior on all of those services. This PE may have a lot of customers on it, which goes against the rule of thumb of minimizing the impact on the starting point, but the focus on this specific use case is to monitor the functionality and discover possible issues to be mitigated on the rest of the network. You should have a strong well-documented backout procedure in these scenarios.

Other factors could influence the starting point of a migration project. It is possible that there is an important customer that needs to be satisfied for their value to the business, and they are willing to accept the risk of impact for a certain upgrade to their service. In most networks, some devices have been there for many years past their lifespan. You could also decide to start your migration on this area of the network as you want to shift traffic and services out of those legacy devices as fast as possible because of the concern that they could just die by themselves and never recover.

Peer Review

It is good practice to have the MOPs checked by at least one other engineer involved in the project. As the saying goes "two heads are better than one". The other engineer may notice missteps, misconfigurations, or missing information on the MOP being checked. That is important feedback for any migration plan.

Service Provider network changes move slowly due to their size, complexity, approval procedure for changes, infrastructure buildouts, and the risk of affecting thousands of customers with just one change. Because of this fact, peer-reviewing every single MOP for a migration project on an SP network might not be feasible. The more peer reviews you complete, the longer it will take the team to finish the migration. This may be okay for some projects that don't demand urgency but for others, it may be best to accept the risk to continue moving forward.

There are multiple options when it comes to peer reviews. One option is to peer review only the template configurations. However, you are risking not seeing all the details of a certain step. Another option is to review only one of the MOPs if they are to be repeated to multiple similar devices. Finally, one last option is to review only major MOPs. These are MOPs that have the highest risk for traffic and customer impact. This serves well for teams with seasoned network engineers that have the experience and have earned the trust of the team to complete most migrations successfully. There is no one right answer here, you must make the determination based upon the complexity of the migration project, the size of the network, the skills of the team members, availability, and timelines of the project.

Implementation

After all the planning and testing have been done, it is now showtime. In preparation for the night of the change, you should coordinate at least one engineer from the team to be rested and ready to pick up any troubleshooting the morning following the change. You also have at least one engineer from the NOC on the call with you for the change to be on the lookout for any alarms or issues in the network due to the changes.

If you are making changes that could make you lose remote connectivity, it is important to ensure you have some type of out-of-band connectivity and/or console access remotely. If that is not present at that location or for the device in question, then you could always coordinate someone to be onsite in case you need console access locally to revert the change or fix an issue. Another option, if supported by the vendor, is to set up a scheduled reboot. This way if you configured something that made you lose connectivity to the box, you know it will reload eventually.

In case you run into an issue, stay calm, you are on a change window. Customers, whether internal or external, should have been made aware of possible downtime. Yes, the idea is to minimize downtime even on a maintenance window. However, it is also important to determine what went wrong in your change and if it is something that needs to be fixed or that will require a modification of the rest of the migration plan. For this reason, you should troubleshoot as quickly as possible while gathering as much detail as possible about the issue.

If it is something a small fix or a misconfiguration that was typed in, then you can just get it fixed and move on. Do not move on to the next step until you have gotten the "okay" from the NOC. If it is an issue that you can't seem to resolve in 30 minutes to an hour at the most, it is something you probably don't want to pursue further. You must decide if you should revert the change at this point. Please remember to log your sessions to continue the investigation later either by the team or with the vendor through a technical problem case.

The NOC is not only there to monitor for alarms and issues. In the same way you have verification steps on your MOPs, the NOC can also verify many different functionalities of the network. They usually have a direct line to the customer call centers for all types of services the company offers. Also, they may have extended visibility into other parts of the network which you may not. Therefore, it is good practice to always verify if things are looking good with the NOC engineer on the call after a major step is completed on the change window.

Finally, after you finish your changes, you should ask the NOC to take one final look around the network and to check with the call center(s) for the service you may be interrupting to see if there are any customers reporting issues. They will have a good idea if services are affected based on the reported issues and the number of customers calling it and at what time. Once the NOC gives the all-clear, you are hands-off and you can go rest.

As part of the change window, you should also plan and execute failure testing. In the testing section of the book, we discussed its important role in determining what works and what doesn't when different failure scenarios occur. However, it is very difficult to mimic the exact behavior of production traffic in the lab. Even if you can simulate huge amounts of traffic, you will probably never be able to account for the different types of traffic seen in an SP network.

In the past, we have seen bugs being triggered by a specific pattern or encoding inside one single packet. Therefore, it is always recommended to do failure testing in your change windows to ensure the traffic fails over as you expect it. This will avoid surprise outages later. There are some network simulation tools that allow you to virtually mimic your network, along with traffic levels, where you can test migrations in it and show you where the traffic will flow for each failure scenario.

Chapter 8
Mergers

One of the most common ways for Service Providers to grow their business is through acquisitions. There are usually two types of acquisitions. One is the acquisition of a section of the network of the SP being acquired. Second is the acquisition of the entire network of the SP being acquired. The two scenarios have different but similar migration and merger life cycles. We will cover the different stepping stones in the life cycles of these network mergers. The process to complete the technical migrations in a merger is the same as discussed in previous chapters. Therefore, not many technical details will be provided here. We will focus on the things to watch out for when doing mergers.

Access

The first step for network engineering after the acquisition is official, is to get access to the network devices of the acquired company/network. Remember, depending on the type of acquisition, you may only be able to request access to the assets under contract. The fastest way to achieve this milestone is by having the acquired company create a Terminal Access Controller Access Control System/Remote Authentication Dial-In User Service/Lightweight Directory Access Protocol (TACACS/RADIUS/LDAP) account for the acquiring company's engineers. Depending on the network, the acquired company may not have these systems set up and the only option is to create a local user account and deploy it to all the contractual devices.

The local user option may take longer if there are many devices in the contract and no automation in place to deploy the configurations quickly. The TACACS/RADIUS/LDAP account creation may also take some time if the acquiring company has many engineers and there are multiple accounts to set up with different permissions. Either way, the first goal is to get access to the network to start documenting it.

As with every network these days, there are public and private networks. It is common to have management IP networks on private IP space. Most acquired companies will push back when requests come in from the acquiring company to connect any equipment to their firewalls. It is therefore a good idea for the acquiring company to deploy its firewalls. This way they can connect to the management network of the acquired network as well as be able to deploy VPN appliances along with jump boxes to give remote access to their engineers.

The acquiring company should also request access to the acquired company's console servers. The request should be for the internal connectivity as well as out-of-band if it exists. When doing mergers, there will be many changes that could be risky, and you should do them through the console connection in case remote access to the equipment is lost.

This might be a bit obvious but it's important to mention when talking about access in acquisitions. It must be specified to the acquired company to update their Access Control Lists (ACLs), if any are present, to allow the acquired company's IPs to Secure Shell (SSH) into the equipment.

There will come a time when you must decide to migrate from their current TACACS/RADIUS/LDAP system to your own as the acquiring company. If you use a different protocol than what they are using, then you could adjust the preference for authentication and authorization by preferring your protocol first. If both companies use the same protocol, then you must replace the configurations and ensure you have connectivity. With this approach, you could set up the local credentials to be preferred before the migration to avoid getting locked out of the box. Remember to update the respective ACLs to allow the TACACS/RADIUS/LDAP server to communicate with the network equipment and vice versa.

Documentation

Documenting the acquired company's network is crucial to putting a migration and merging plan together. You need to identify their architecture, protocols, network devices, software versions, services, and any special features. Not all SP networks are documented in detail. Some are not documented at all. It is your responsibility as the network lead to ensure the network is documented. You shouldn't develop or approve a migration/merger plan without this documentation.

If the network is documented well, then it is a manner of verifying it is up to date and has no gaps. If there is little to no documentation, then it is best to document as much as possible by logging into the network devices or using a mapping tool if possible. Once you have mapped out what you can, it is time to work with the network engineering team of the acquired company to highlight the hidden functionalities that may not be identified in the first round of documentation. Most SP networks have temporary fixes or architectures that were meant to be temporary but were never changed or cleaned up. The earlier in the process you identify these, the lower the risk when you go to migrate/merge.

IP Planning

One crucial part of the documentation is IP planning. Most SP networks utilize some networks of the RFC1918 space.

Because of the scale of SP networks, it is likely that the acquiring company and the company being acquired are actively utilizing some of the same network blocks on their respective networks. This is known as having IPv4 conflicts.

It is ideal to identify the current IPv4 conflicts early in the process as well as note what each private block is being used for and where in the network. Sometimes these private blocks might be inside a Virtual Routing and Forwarding (VRF) instance, a scenario for which you should keep an eye out. There are three options when it comes to IPv4 conflicts.

1. Move to IPv6 addressing. On a migration/merger project, that may not be feasible for either side due to timelines and technology constraints.
2. Replace the current private blocks with conflicts in the acquired company with unused private blocks from the acquiring company.
3. Use Network Address Translation (NAT) to integrate both networks.

It should be noted that both companies on the acquisition will keep operating Business As Usual (BAU) during the merger process. Sales and activations must continue to financially support each business. It is therefore essential that there is good communication between both companies when deploying new networks from the private IP space in the acquired company. This is to avoid deploying even more IPv4 conflicts without either company knowing about it until an issue arises.

Project Tracking

On almost all network acquisitions, there will be a project management team to ensure all the tasks are being tracked and completed in time. Also, to ensure proper decisions are made when issues arise. Be specific with the project manager when you highlight your migration/merger plan. They can't help you if there are missing pieces to the puzzle. Communicate effectively and often with them and escalate issues when necessary to ensure you stick to your schedule and can meet your target timelines.

Project managers will also ensure proper collaboration is taking place and you have everything you need to continue making progress. Highlight any dependencies with other teams on your migration/merger plan. Identify risks properly and do not commit to dates unless you are sure you can meet them. Sometimes people feel pressured to commit to a date because they are in a meeting with the higher-ups. Don't do that. Remember that their success is contingent on your success. It is okay to push back. Just be honest about when you can complete your tasks. As the saying goes "under promise and over deliver."

Merger Technical Plan

Complete Network Acquisition

As mentioned before, there are two scenarios when it comes to network mergers. We will first cover the scenario of the acquisition of the entire network of a certain SP. This is the simpler of the two scenarios as you are taking over all the assets and will have full control of the network once it is official. Carving out control of part of a network is never easy, as you must identify what is coming over and what is not as well as having to work with the acquired company constantly to ensure the services are properly migrated.

When it comes to technical planning of the migration and merger of an entire network, assuming all the steps before this section have been completed, the first overall objective for core routing would likely be to migrate traffic off their transit links (if any are present). If the acquired network is a strict MPLS private network with L2VPN or L3VPN customers, then this step can be skipped, and you can move forward with the technical planning of merging the two networks. To migrate traffic into your transit connections, you will need to figure out the underlay and overlay networks needed to achieve this as well as the infrastructure buildout necessary.

Sometimes you may get lucky, and the assets inherited are in the same major colocation facility. All you need to do is run a couple of cross-connects to achieve this milestone. Other times, you will need to figure out the merging points of the networks as well as design how to interact with their routing architecture to get prefixes advertised in and out from your transit connections. You should ask for the contractual end date to get the traffic moved off. Usually is six months to a year under contract but it varies by acquisition.

The immediate critical contractual milestones are taking over control of the access, operations, and monitoring of the network, as well as moving traffic off their transit circuits. Once this is completed, the real merger process usually begins. When merging over an entire network, you must analyze its architecture and determine what to keep and what to replace. This includes the underlay network and the overlay network. You could keep their IGP as is because it is the same as yours or because it fits the long-term architecture plan.

One thing to note if keeping the IGP is to make sure to standardize it. Adjust the timers to the same ones you use, remove redistribution, move the links to point-to-point, etc. Applying new IPs to the loopbacks for standardization is another option. More than likely you will be owning the IPs already deployed as part of the acquisition, but you may want to re-IP for standardization purposes. The same applies to the MPLS and BGP architectures if present in the acquired network.

Another key decision before starting the merger is whether to keep the network devices and logical architecture currently in place or replace them with your standard and approved ones. Sometimes there is a hurry in merging the networks due to internal systems needing to talk to the newly acquired network, expiring leased circuits, consistent instability in the network, or some other reason. If that is the case, you may not have the time to replace the equipment and you must push forward with the merger. That is completely fine. If a replacement of the equipment is in order, it can be done before, during, or after the merger.

Doing it during the merger is probably the best of both worlds in terms of minimizing downtime and timeline to completion. This is because you can deploy the new equipment with the new physical infrastructure and just move connections as the merger happens. This way you can achieve the physical and logical merger sort of at once. It is a neat approach as it is somewhat greenfield but not really. Doing the replacement before can be recommended in case immediate stabilization is needed. Doing it after is probably the safest as you can greatly minimize the risk of downtime by taking as long as you need on the planning, designing, and execution just like any other migration. It will all come down to how much risk management is willing to take to make the replacement happen during any strict timelines they may have.

When you have the technical merger low-level plan in place, it is suggested to constantly look at your MOPs. You should look at each service you are migrating and visualize the traffic end-to-end for it. Make sure your configurations cover traffic reachability both ways. The more you look at them, the more you will be able to find out if something is missing. Don't be discouraged if you keep finding missing configurations, this is normal when doing mergers due to the complexity of mixing and matching protocol and architectural behaviors.

Partial Network Acquisition

Acquiring a partial network of an SP is complicated for both sides. Not only in the technical sense but contractually. When merging these two networks, the same major milestones apply as with the scenario above. However, it will likely take longer to achieve them as you need to constantly work with the acquired company to deploy equipment and changes since there are clear demarcation points as per the contract on who owns what and where.

Many SPs provide broadband or Fiber-to-the-Home (FTTH) services. When acquiring a network that is providing such services, it is recommended to wait for the broadband/Passive Optical Network (PON) and provisioning teams to complete their migration. This is because there are many moving pieces to such a migration in terms of provisioning servers, voice services, and DNS to name a few. But the main reason to wait on this migration is IP network changes. During most partial network acquisitions, there will be a transfer of IP blocks. There will be blocks that the acquired company will keep and some they will transfer over. Because of this, the migration of the residential services becomes a shell game of IPs. It constantly changes at different points of the migration. Therefore, it is best to wait until they reach a final deployment in terms of IPs before executing the core routing migration and merger to avoid going back to changes already made.

If the acquired network provides L2VPN and/or L3VPN services, it is important to document all the endpoints of each of those customers. If all the endpoints for a customer are in the footprint of the acquired network, then you can follow the merger transition plan as discussed. If one or more of the endpoints are outside the acquired footprint, then you have a decision to make along with the company the acquisition was made from.

The first decision is who is keeping the customer, as it will now have endpoints in two different entities. Depending on who is keeping the customer contractually, the next step is to decide whether the company keeping the customer will be buying Network-to-Network Interfaces (NNIs) from the other company for the customer endpoints outside its footprint that are still on the footprint of the company losing the customer. Another option is to set up one of the MPLS Inter-AS options (a, b, c, or d). Technical details of these options are outside the scope of this book but there are many white papers and trainings out there around this topic.

In every network project, you will find skeletons. Almost every network has temporary fixes in their network that were never cleaned up. Some of these fixes have been in place for years, and it is difficult for any single engineer to remember all of them off the top of his or her head during the network discovery conversations with your team. Therefore, that first merger change is extremely important for finding these skeletons and how to fix them. Sometimes you find the skeletons before this change and you may see minimal outages. Other times, you will have massive outages until the fix for these unknown issues is applied.

The rule of thumb when it comes to fixing these issues, especially during a massive outage, is to find the quickest and simplest solution to bring the service back to customers. For this first change, you should make sure you have proper support from the NOC and other engineers from your team as well as engineers ready to go early in the morning in case troubleshooting needs to be done. When these outages occur, focus on one issue at a time. Usually, the issue with the greatest number of customers down is picked or some VIP customers that are important to the business. You should keep this in mind and verify your changes on this first change as thoroughly as possible.

When making network changes during such complex mergers, it is recommended you use an out-of-band console connection or a local user account with the proper privileges. This is to avoid getting yourself locked or unable to make configurations because the network device can no longer reach the authentication and authorization server due to an unforeseen reaction to a change you just made.

To ensure the success of any partial network merger, you should build great communication channels with the acquired company's network engineering team. Remember that your team will only have access to a portion of their network but will still need to interact with the rest of it until the split-off stage. You will need their assistance with configurations, documentation, and changes in other parts of the network to which you will not have access. You should communicate with them consistently during the entire process and keep them up to date on the progress. You can also push them a bit to get you the information you need if you think you will be falling behind your timelines.

Some companies have silos of teams and others have a better cohesive team that has their hand on many cookie jars. You will need to learn to work with both. Have a clear point of contact list for each service, area of the network, or technology depending on if and how the team silos are operated. Getting through this type of merger successfully is as important to your team as it is to theirs, collaborate efficiently.

Technical Caveats

There are several technical aspects to consider when doing a merger. This is apart from the major one, which is IPv4 conflicts. They are mentioned here for you to keep in consideration when coming up with the technical plan for the merger as well as to offer some ideas on how to integrate or replace them with your long-term architecture.

IGP

The first major obstacle will be the integration of the IGP architecture. In the SP world, you will mainly see either OSPF or ISIS as the IGP in the network. Both can be done flat or designed hierarchically via areas in the case of OSPF and via levels in the case of ISIS. When merging such IGPs, first you must decide if you are going to keep them or not. If you are running a unified MPLS architecture or some other type of architecture where areas of the network are silos in their own IGP islands, then you can skip this section. Moving on with the subject. If your production network is not running the same protocol, then you may not be inclined to keep their IGP and gradually merge it with the one in your production network. However, that may not always be the case. Depending on the acquisition, your company might be merging with a bigger more structured network.

The IGP protocols might be different but, in this case, it is less effort to replace your IGP and merge it with the bigger network. This is very rarely the case where a company acquires a bigger one, but it is not out of the picture depending on the financial health of each business. If your company is acquiring a smaller network, then it is recommended you follow your standard IGP deployment and replace the one in the acquired company if they are different protocols. However, you could also keep the IGPs different and do redistribution or follow a unified MPLS architecture if it calls for it. Details on which IGP is better for scale are beyond the scope of this book, but there are well-known network segmentation architectures for SP networks that can be scaled to thousands of nodes with either ISIS or OSPF.

If the case is that the acquired company shares the same IGP protocol, then a careful thought process must go into comparing both their current IGP architecture to the one you have in your production network. Some basic questions you should answer initially are:

1. Are they using a flat IGP, or is it split into areas/levels?
2. If they are using areas/levels, where are the Area Border Routers (ABRs)/Level1-2 routers?
3. If OSPF areas are used, what are the area types?
4. If ISIS levels are used, are there any routes leaked from level-2 into level-1? If so, where?
5. Is there a default route being injected by the IGP somewhere?
6. Is there any redistribution being utilized?

7. Is there any use of distribute lists anywhere in the IGP?

Once you have this information, you can make the proper IGP merger decision. Whatever the final decision is on which architecture you will adopt, it is best to make the changes on either network before merging them. For example, let's say they have a network split into OSPF areas, but you want to integrate it into your own hierarchical OSPF architecture under a new area without Area 0 on that network. It is recommended to focus entirely on building that area in the acquired company before integrating it into your OSPF architecture. You can do this in two ways, you can build the area on top of the current deployment, or you can slowly change the areas on that network to the new area.

There are a few things to keep in mind when following these types of IGP buildups. You should build the new area continuously. This means the new area always stays connected. If you build one side of the network in the new area and then build out the other side of the network in that same area, these two sides will not be able to talk to each other as per OSPF standard rules. Also, watch out for what IP the vendors use for the Link-State Advertisement (LSA)/Link-State Packet (LSP) originator when building areas on top of each other. You may lose some connectivity due to this issue where the router is using an originator ID it has no reachability to.

When it is time to merge the networks, it is highly recommended (if possible) to use a router that has the capability of using a "commit confirm" mechanism. Meaning the configuration is committed into the running configuration for a certain amount of time and then it rolls it back after that time expires. This way if a major outage is created due to the IGP merge and you lose access to the router, the outage will be minimized to only the amount of time you specified.

MPLS

The term MPLS is used interchangeably to specify a network that can transport different services using some type of ID before the layer 3 routing whether it would be a label or a segment ID. When integrating networks, you must first identify the type of label switching system the acquired company is utilizing. You may see MPLS LDP, MPLS Traffic Engineering (MPLS-TE) tunnels, MPLS LDP over Resource Reservation Protocol (RSVP)-TE, or Segment Routing (SR). You also need to identify if they are using any type of fast rerouting mechanisms such as MPLS-TE Fast-Reroute (FRR) or SR Topology-Independent Loop-Free Alternate (TI-LFA).

If the companies are utilizing the same mechanism, then it is of course simpler to integrate, but there are still some caveats to look out for. Let's go over each scenario. With MPLS LDP, it is probably the easiest to integrate and deploy because of the simplicity of its configuration. It is also a well-known and deployed protocol that most SP engineers are familiar with. With LDP over RSVP, you need to find the merger point where the LDP is tunneled over RSVP. You can then identify your merger points with theirs and come up with an integration plan that works. With MPLS-TE, it is similar to MPLS LDP with the major difference that the configuration is more extensive.

Another caveat to be on the lookout for MPLS-TE on links connecting two different IGPs, there are extra configurations to be noted there. If MPLS-TE FRR is being utilized on either network, considerations must be taken into whether those FRR tunnels will be extended/integrated and where. With SR, the main caveat is the SR global block as it should be the same across all devices in that SR domain. If either network utilizes TI-LFA, the only concern would be vendor support. TI-LFA doesn't require the careful planning that FRR needs as it runs the tunnel calculations automatically once it is configured. FRR can do this as well with one of its features but here we are assuming the tunnels are manually configured.

BGP

One of the things to watch out for when merging two networks regarding BGP is route reflector architecture and its integration. Route reflector location is important if the "additional path" or "optimal route reflector" feature is not used because the location will determine what is the best path and therefore what path the clients get. Some vendors might support the use of multiple routing instances with different ASNs on the same router. Other vendors only support one ASN system-wide. That is also a consideration to be taken into account when merging the RR architectures as some devices on either network could have BGP sessions on different ASNs and might require an ASN consolidation or the use of the "Local-AS" feature.

From the architecture standpoint, the main concern is the difference between a flat RR architecture and a hierarchical one and how to integrate them. One approach is to pick the top-level RRs of the hierarchy and build sessions to a couple of the RRs in the flat architecture.

The last thing to note is features and address family differences. Some RR features might not be too impacting to traffic like soft-reconfiguration/route-refresh, but others might greatly impact the way the traffic flows across the network, and others might be highly important to protecting loops like using cluster-ids. You should document them and come up with a plan for each. The address families utilized in each network might be different. In the initial stages of the merger, the recommended practice is to build sessions if needed for those address families not on your network to maintain customer services. For example, let's say the acquired company provides L3VPN services but you don't. In the RR integration, you should build VPNv4 address family sessions to avoid interrupting the L3VPN service of those customers, even though you don't have any in your network.

Reverse Path Forwarding (RPF)

Each SP will have its transit point connections to the Internet/Default Free Zone. When doing a merger and taking over IP space from the acquired company, it is likely that during the migration or merge both networks will be advertising the same IP prefix(es) out of their transit connections. There is no issue with this, and you can control the traffic using AS-path prepend or local preference to engineer the traffic flow.

However, you may see some asymmetric routing during this stage of the process as maybe the acquired company is advertising a smaller block, or someone misconfigured the BGP route policies. Usually, asymmetric routing is not a major concern for residential services if the latency is not extensive. The problem arises if the acquired network utilizes RPF on their transit connections, and you are not aware of it. RPF will drop asymmetric routing in certain cases. The dropping could vary depending on whether "loose" or "strict" mode is used. This is something you should be on the lookout for when doing your merger planning.

Chapter 9
Migration Variables

Equipment Ordering

Before we get into some detail about ordering from vendors, please try to never order any equipment for migrations before testing is complete. This is not always the case as sometimes projects require very short completion timelines and it may not be possible to onboard a new vendor due to legal or purchasing department processes. However, it is important to remind you to always test to avoid further issues and added complexity to the network.

You can see that building infrastructure would require many teams working together as well as vendors. You need to reach out to the vendors and get quotes for new orders. You need to know exactly how many of each kind of equipment you are going to order (don't forget about ordering spares). You need to know where things are going to be shipped to and who is your point of contact for delivery. You also need to check the budget to know how much you can order from each vendor. It is possible you may not have enough budget to order from the vendor you want and you may need to shift focus to other possible equipment that is more affordable and can do the job. Finally, you need to know from your vendors whether the equipment will be delivered in time to meet your timelines, or maybe you need to go with another vendor or get creative.

Resources

Very few migration projects get completed by a single person. Most of the time there is a team either on stage or behind the scenes that play a crucial role in the success of the migration. Some of these teams have already been mentioned but their value is important enough to warrant a section of their own.

The engineering team is usually composed of junior to senior engineers. Depending on the company, these engineers might be segregated into specific areas of the network like core routing, commercial, broadband, PON, access, etc. You must have a firm grasp of the skill of each engineer to be able to assign them specific tasks within the migration plan. The best way to learn the skills of engineers is by collaborating with them on different projects.

It is always a good idea to identify the weaknesses and strengths of the skills of each engineer. Your role as the lead for the migration plan is to bring in engineers as you see fit to help execute the migration successfully. You may need to bring an engineer from outside of your direct team because that person is strong in a specific skill. Let's say you are doing a core routing migration but you need someone to develop the scripts quickly to deploy configurations to 500 devices. You may not have that skillset in your direct team but you know someone else who does.

Or let's say you are migrating routes from OSPF to BGP on broadband routers. You may want to have a broadband engineer with you to verify the modems come back up and to help troubleshoot anything that is needed on the broadband side. Being able to identify the right resource from the internal team will save you time when doing such migrations. Go out there and collaborate.

There may be migration projects that are just too big and/or complex to complete in the required timeline based on the skillset and availability of the engineering team. You should identify if this is the case for the project you are working on and decide whether to bring outside resources to help out. These resources are called contractors. There are multiple companies out there that can provide engineering resources for many different tasks. It could be for racking and stacking, migration strategy design, building MOPs, executing MOPs, staging, and many other others. No matter the task you need help with, there will be someone out there willing to help out …. for the right price of course.

The best practice when dealing with contractors is to be very specific about the task you are seeking help with. They usually charge by the hour. The grayer areas you leave on your project/task description, the more hours they will put into the proposal quote as they have unknowns they need to account for. It is also important to negotiate. The consulting firm will always add a buffer on the proposal quote to account for possible delays or unknown variables. They always have wiggle room to negotiate, be mindful of that. Finally, you have the option of contracting out the entire migration project if the team is too busy with something else more crucial with super aggressive timelines. There are great contractors and there are average ones. Try them out on simpler migration projects and see which one fits your needs and performs the best. You can mix and match them based on what you see from them on different types of projects.

We have already talked about the NOC and its role in a migration plan. It is good to mention that they should also receive a short and high-level explanation of what the migration plan is. There should be constant communication with the NOC leads to coordinate maintenance windows so they can alert customers in advance of when they might see their services affected. The NOC engineers can also be leveraged to complete certain tasks of the migration plan. They have their own set of skills and most (if not all) are willing to help with things outside their direct responsibilities. Leverage them accordingly, they are an invaluable resource.

Team Training

On migration projects, not all team members may be involved either due to specific skills, aggressive timelines, or organizational silos. Because of this, it is important to provide proper training to everyone who will be responsible for operating, troubleshooting, and implementing future changes on the network after the migration project. The training and its details should be tailored for the audience. If the audience is the NOC team, then the focus should be on troubleshooting, finding the root cause, and fixing the issue. If the audience is the implementation team, then the focus should be on turning up services. You can always cover everything under the same training to cover all aspects. That is recommended, as that way everyone has visibility to the entire life cycle of the functionality. It is also a great idea to record these training sessions to avoid having to repeat them when new team members join the different teams.

Emergency Changes

In most SP companies, there are committee approval meetings where changes for the next week are approved or rescheduled. There will be times when changes need to happen before they even go through the committee approval. This could be to fix a major issue or to meet aggressive timelines of a certain phase of the project. Regardless of the reason, you should have a specific plan for the change.

Because the change might be happening in a short amount of time, you probably will not have enough time to put a ton of detail into the change. You should keep the change as simple and short as possible and with as few configurations as possible. This is to minimize the risk of the change as it will be happening outside the normal planning timeframe, and you may not be able to think of every possible outcome since it may not have gone through testing.

For emergency changes just focus on the problem or the task at hand. Nothing more, nothing less. These types of changes should also have rollback procedures in place. Always check with the NOC and implementation teams to ensure they are aware of possible last-minute disruption of services whether internal or external.

Creative Network Engineering

Let's say the equipment you had in mind will not be delivered in time. The migration now takes a turn. As a team, you must decide what to do. You can figure out a way to utilize the current infrastructure by coming up with another set of features or protocols that are supported by the current equipment. You can decide to put a temporary migration plan in place while you wait for equipment. Finally, you can decide to extend the timelines and follow a long-term plan. Let's also assume you get a set of requirements out of nowhere that needs to be implemented in a very short amount of time. You must think fast and come up with the simplest solution that satisfies the requirements while keeping complexity at a minimum.

I call this creativeness "magic" because as an engineer your job is to see a problem and come up with the best solution possible. The solution might not be super clear to everyone at that time, but it is your responsibility to explain it in detail and justify the solution along with your short-term and long-term plan. When these situations arise, most of the time the short-term solution will vary from the long-term one. That is okay, it is part of being a creative engineer in a pressured environment. Always keep learning about your network and new technologies that are coming up, so you can keep those magic tricks in your back pocket.

Keep networking with network engineers to ensure your magic powers never wear off.

Index

6

6 Provider Edge
   (6PE) 15

A

Access Control Lists
   (ACLs) 63
Area Border Routers
   (ABRs) 75
Autonomous System Numbers
   (ASNs) 35

B

Bidirectional Forwarding Detection
   (BFD) 21
Border Gateway Protocol
   (BGP) 4
Business As Usual
   (BAU) 65

C

Capital Expenditures
 (CAPEX) 4

D

Data Center
 (DC) 20
Direct Internet Access
 (DIA) 28
Domain Name System
 (DNS) 24

E

Ethernet Virtual Private Network
 (EVPN) 6
external BGP
 (eBGP) 38

F

Fast-Reroute
 (FRR) 77
Fiber-to-the-Home
 (FTTH) 70

G

Generic Routing Encapsulation
 (GRE) 22

I

Interior Gateway Protocol
    (IGP) 15
Intermediate System to Intermediate System
    (ISIS) 17
internal BGP
    (iBGP) 38

L

Label Distribution Protocol
    (LDP) 50
Layer 2 VPN
    (L2VPN) 6
Layer 3 VPNs
    (L3VPN) 56
Link-State Advertisement
    (LSA) 76
Link-State Packet
    (LSP) 76

M

Methods of Procedures
    (MOPs) 29
MPLS traffic engineering
    (MPLS-TE) 77
Multicast Distribution Tree
    (MDT) 22

multicast VPN
    (mVPN) 22
Multi-Protocol Label Switching
    (MPLS) 15
Multi-System Operator
    (MSO) 24

N

Network Address Translation
    (NAT) 65
Network Operations Center
    (NOC) 31
Network-to-Network Interfaces
    (NNIs) 71

O

Open Shortest Path First
    (OSPF) 17
Operating Expenses
    (OPEX) 4

P

Passive Optical Network
    (PON) 70
Points of Presence
    (POPs) 11

point-to-point
  (P2P) 28
Protocol Independent Multicast
  (PIM) 21
Provider Edge
  (PE) 55

R

Resource Reservation Protocol
  (RSVP) 77
Return on Investment
  (ROI) 4
Reverse Path Forwarding
  (RPF) 80
Route Reflectors
  (RRs) 16

S

Secure Shell
  (SSH) 63
Segment Routing
  (SR) 77
Service Provider
  (SP) 1
SR Topology-Independent Loop-Free Alternate
  (TI-LFA) 77

T

Terminal Access Controller Access Control System/Remote Authentication Dial-In User Service/Lightweight Directory Access Protocol
    (TACACS/RADIUS/LDAP) 62
Trivial File Transfer Protocol
    (TFTP) 20

V

Vice President
    (VP) 3
Virtual Local Area Network
    (VLAN) 28
Virtual Routing and Forwarding
    (VRF) 65

www.ingramcontent.com/pod-product-compliance
Lightning Source LLC
Chambersburg PA
CBHW070245220526
45465CB00004B/1528